SpringerBriefs in Applied Sciences and Technology

More information about this series at http://www.springer.com/series/8884

Norazrin Azwani Ahmad
Masine Md. Tap · Ardiyansyah Syahrom
Jafri Mohd Rohani

Quantitative and Qualitative Factors that Leads to Slip and Fall Incidents

Springer

Norazrin Azwani Ahmad
Department of Materials, Manufacturing
and Industrial, Faculty of Mechanical
Engineering
Universiti Teknologi Malaysia
Johor Bahru, Johor
Malaysia

Masine Md. Tap
Department of Materials, Manufacturing
and Industrial, Faculty of Mechanical
Engineering
Universiti Teknologi Malaysia
Johor Bahru, Johor
Malaysia

Ardiyansyah Syahrom
Sport Innovation and Technology Centre
(SITC), Institute Human Centred
Engineering (IHCE)
Universiti Teknologi Malaysia
Johor Bahru, Johor
Malaysia

Jafri Mohd Rohani
Department of Materials, Manufacturing
and Industrial, Faculty of Mechanical
Engineering
Universiti Teknologi Malaysia
Johor Bahru, Johor
Malaysia

ISSN 2191-530X   ISSN 2191-5318   (electronic)
SpringerBriefs in Applied Sciences and Technology
ISBN 978-981-10-3285-1   ISBN 978-981-10-3286-8   (eBook)
DOI 10.1007/978-981-10-3286-8

Library of Congress Control Number: 2016958497

Printed on acid-free paper

This Springer imprint is published by Springer Nature
The registered company is Springer Nature Singapore Pte Ltd.
The registered company address is: 152 Beach Road, #22-06/08 Gateway East, Singapore 189721, Singapore

This book is dedicated to the man and woman who raised me and making me be who I am; they are my parents. It is also dedicated to the man who is supporting me all the way; he is my husband. Not forgotten, to the little boy who makes me a wonderful woman; he is my lovely son.

Preface

Workplace-related accidents have continued to rise annually in Malaysia. Many factors combine to change the work environment significantly such as globalization, migration, demographic change, advancing family structures and the effect of financial crisis which have great impact on worker's safety and health. A total of 3,263,058 days were spent for medical leaves by the victims of occupational accidents due to various types of injuries in year 2014 and the numbers are continuously increasing. For this reason, occupational accidents should be avoided since they will adversely affect the individuals or the companies involved. There is a need to study slip-and-fall areas since it can represent as a personal injury case. Slip-and-fall can be found in almost all types of working environment. In human perception issue, slippage of the foot on the floor is due to the misperception of risk. This is because they cannot maintain their body balance and adjust gait pattern appropriately. As a result, he or she will experience pain on certain parts of the body. In worse cases the incident would have caused permanent disability (PD) or death.

Johor Bahru, Malaysia

Norazrin Azwani Ahmad
Masine Md. Tap
Ardiyansyah Syahrom
Jafri Mohd Rohani

Acknowledgement

This project was sponsored by the Kementerian Pendidikan Malaysia (KPM) through grant scheme (R.J130000.7809.4F355). The authors would also like to thank the Research Management Centre, Universiti Teknologi Malaysia, for managing the project.

Contents

Abstract

Fall has been identified as one of the common types of accidents that occurred in manufacturing industry. Fall-related accidents have not only affected the victims' health but also are very costly. Significant gait changes are made when a person is aware of the potential risk of slipping. This kind of perception will make a person walk more carefully in the face of possible fall accidents, thus reducing its possibility to happen. However, a wrong perception of floor slipperiness may cause a person with inappropriate gait pattern a slippage of the foot on the floor which may lead to fall accidents. The main objectives of this study are to establish factors that lead to slip and fall incidents, to establish relationship between coefficient of friction (COF) with floor slipperiness and floor roughness, and to establish the human perception of slipperiness through measured COF. Epidemiology approach is used through questionnaire survey for the identification of potential risk factors in manufacturing workers population. Tribology approach is used in experiment to relate the interaction between contaminants, floor and footwear materials. Human-centered approach is also used through questionnaire survey to addresses slipperiness measurement by human perception. The findings show that perception of risk is the main factor leading to slip-and-fall among manufacturing workers. In addition, it was concluded that friction was significantly affected by footwear materials, type of floors and the presence of contaminants on the floor surface. The COF is also related to surface roughness of the floor types. It was discovered that the human perception for slipperiness was consistent with the various contaminant conditions represented by COF. However, human perception for slipperiness was not consistent with the COF of floor types. This inconsistency is significant in assisting floor designer in designing safe floor that gives the right human perception. It also helps the employers to form regulation on selection of a suitable floor and footwear material.

Abstract

Chapter 1
Introduction

Abstract This chapter reviews the elements in slip and fall system. The phenomena of slip and fall are briefly described before presenting the quantitative and qualitative method to discover slip and fall incidents. The slip and fall overview and background are presented.

Every year, workplace-related accidents continued to rise significantly in Malaysia. In his speech during the International Commemoration Day for Dead and Injured Workers 2012, the Vice President of the Malaysian Trades Union Congress mentioned that work-place-related accidents in Malaysia are considered high if taking into account the number of workers in this country. Over the last seventeen years, a staggering 2 million people have died from workplace-related accidents world wide while another 1.2 million peoples were injured and more than 160 million workers fell ill due to unsafe, unhealthy, and unsuitable workplace [1].

A total of 63,331 accident cases were reported in 2014 [2]. In 2013, a total of 63,557 accident cases [3] were reported, a marked increase of 2005 cases or 3.26% from the number of cases in 2012 [4]. More than half (57.43%) of these accidents occurred at workplace. In addition, the Department of Occupational Safety and Health (DOSH) recorded 1248 workplace accidents in ten sectors in the first five months of 2013 [5]. Between January and June the following year, 1512 workplace accidents were recorded [6]. To be specific, slip-and-fall incidents recorded 17,803 of total cases and the person falling from the same level recorded 8963 cases in year 2014 [7].

Reducing occupational accidents is very important because a spotless occupational and safety record would be good for the country's image and foreign investment initiatives [8]. This is in line with the focus of national budget which is to increase the investment activities as the local and international companies become confident with a country's level of concern for labour.

Since 1970s, there have been many factors being considered to tackle this problem. Several studies were done to measure the slip using various key factors. Initially, the measurement of slip was introduced and performed using biomechanics approaches [19].

© The Author(s) 2017
N.A. Ahmad et al., *Quantitative and Qualitative Factors that Leads to Slip and Fall Incidents*, SpringerBriefs in Applied Sciences and Technology, DOI 10.1007/978-981-10-3286-8_1

1.1   Occupational Accidents and Injuries

Fostering an occupational safety and health culture and accident-free workplace environment in Malaysia is still a challenge [9]. In January 2016, 285 cases were recorded [10]. As of August 2015, the total number of occupational accidents by sector were reported to be 2025 cases [11], while in 2014 the total number of occupational accidents were 1667 cases [6]. These statistics show a worrying trend, hence employers need to improve their workers' safety awareness especially among those who are exposed to hazards at workplace.

A total of 41,404 days were spent for medical leaves by the victims of occupational accidents due to various types of injuries in the year 2012 [4]. This figure increased to 3263,058 days in the year 2014 [7]. While medical leaves can serve as a basis to assess injury severity, they are also the cause of the loss of productivity. For this reason, occupational accidents should be avoided because they will adversely affect the individuals or the companies involved. Furthermore, some of the medical costs have to be certified by an individual, a company, or an insurance agency [12]. In the case of fall-of-person accidents, the incidents recorded 17.2% (RM 325,612.27) from the total cost of injury [4, 13]. This implies that fall-related accidents have not only affected the victims' health but also their spending.

Therefore, there is a need to study the slip-and-fall area based on the above mentioned point. From human perception issue, slippage of the foot on the floor is due to the misperception of risk because of the failure to maintain body balance and adjust gait pattern appropriately. As a result, he or she will experience pain on certain parts of the body. In worse cases, the incident would have caused permanent disability (PD) or death. All of these adversaries pose severe impact on the victims and the employers.

1.1.1   Manufacturing Industry

To be an industrialised economy by the year 2020 [14], Malaysia has chosen industrialisation as an integral part in its development strategies since 1960s. Manufacturing sector has become one of the important backbones of Malaysia's economy as it creates huge employment opportunities in the country. From the record, manufacturing sector represented only 9.4% of the total employment in the beginning [15] but then increased to 27.8% in the year 2010 [16].

Manufacturing sector can be considered as the most dangerous industry in Malaysia [11] because having multiple activities at one manufacturing site increases the risk of an accident to occur. Thus, efforts have been taken by researchers, organisations, and government agencies in the field of health and safety to propose various improvements to prevent this matter. However, the involvement of the public and private sectors to comply with Malaysia's standards of occupational safety and health (OSH) is needed [9].

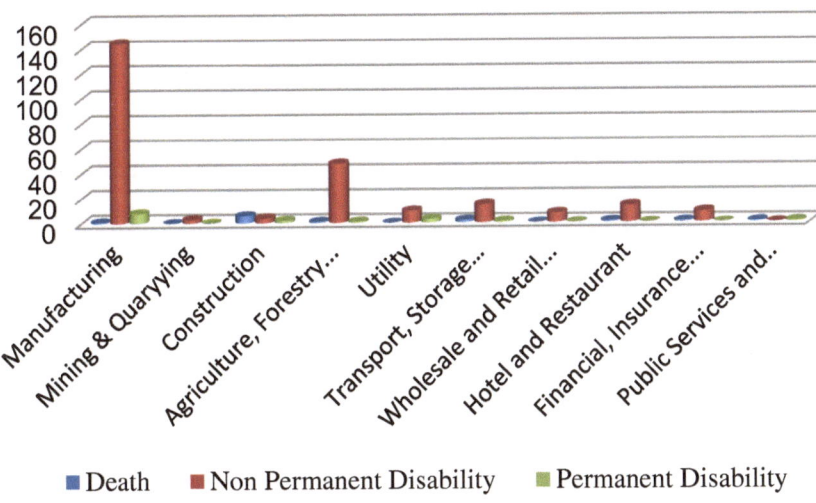

Fig. 1.1 Industrial accidents reported by sectors in Malaysia [10]

Figure 1.1 shows the statistics published by the Department of Occupational Safety and Health for occupational accidents by sector for January 2016 [10]. The figure demonstrates that, manufacturing sector has the highest number occupational accidents compared with other sectors. Following the manufacturing sector are the agricultural, forestry, logging, fishing and transport, storage and communication sectors. Besides that, other cases reported from the concerned sectors were 1 case of death, 145 cases of victims facing non-permanent disabilities, and 8 victims facing permanent disabilities. Due to manufacturing variability as well as complex tasks and activities, this sector is facing the highest number of occupational accidents. Thus, fundamental changes especially in terms pf the way organisations act on the safety issue may provide a safe and healthy working environment.

1.2 Types of Accidents

There are many types of accidents that usually occur at the workplace, and fall has been ranked as the third common type of accidents [17]. Data from the Department of Occupational Safety and Health are adapted into Pareto charts in Figs. 1.2 and 1.3. Commonly, accidents occur after victims get caught in or between objects, or by stepping on or come into contact with an object. Fall-of-person recorded 456 cases in the year 2012, indicating 1854 in cumulative count and 75.3% in

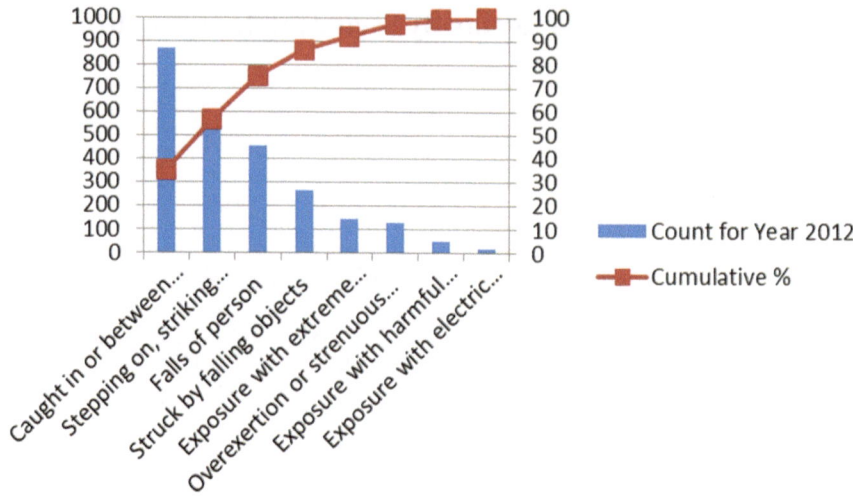

Fig. 1.2 Types of accidents for the year 2012 [17]

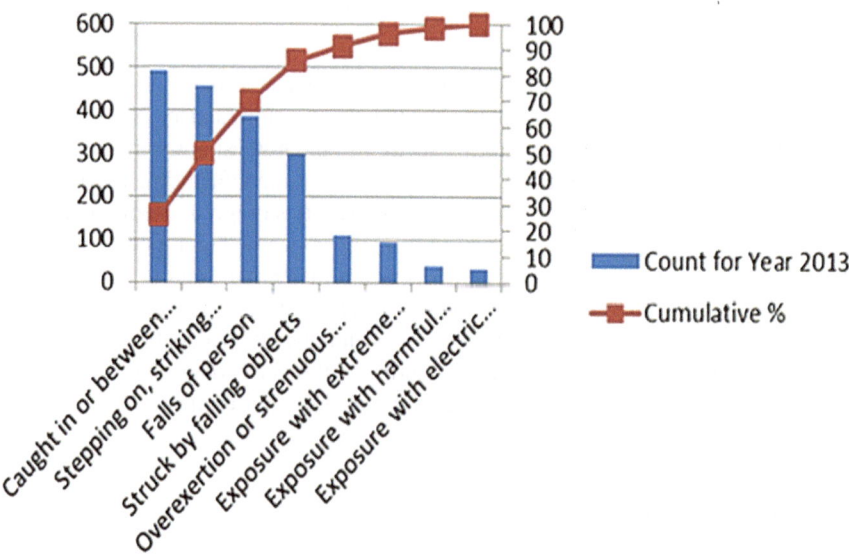

Fig. 1.3 Types of accidents for the year 2013 [17]

cumulative percentage. Subsequently, the following year recorded 385 cases of fall, with 1333 in cumulative count and 70% in cumulative percentage. The data show the importance of establishing a security-conscious culture and environment. This is aligned with a statement made by the chairman of Malaysia National Institute of Occupational Safety and Health [9].

1.3 Types of Fall

There are two types of fall namely elevated fall and same-level fall [18]. Elevated fall is the kind of fall that has a low frequency of occurrence but high severity. On the other hand, same-level fall has a high frequency of occurrence but low severity. In a general population of work-related injuries, same-level fall leads the cases of fall, and slipping is a predominant cause of fall.

Slip can cause a person to experience either elevated fall or same-level fall. In this case, each slip-and-fall incident has its own context in different populations. For example, adults typically experience fall at the workplace or during sports activities. Meanwhile, domestic and sports are two areas where children usually get involved in fall, and older people are experiencing it in the areas of domestic and care.

1.3.1 Elevated Fall

According to a previous study in agricultural sector in Florida, 17% of all serious injuries were from the elevated falls and 8% were from the same-level falls [18]. However, there are also a significant number of falls from vehicles and equipment, loading docks, buildings, and other structures. They are:

(i) Falls from the ladder
   Usually, ladder-related injuries occur when a person attempts to reach too far left or too far right. When working on a ladder, the person's belt buckle should never expand over the rails on both sides because reaching further may cause the ladder to slide in the inverse direction.
(ii) Falls from vehicles and equipment
   Death or serious injury is a frequent consequence of extra riders tumbling from tractor, equipment, or the bed of a truck. Too many injuries happen during the simple process of getting in and out of truck, on or off tractor, machinery, and so on. It is important to keep the steps clean and dry to avoid falls from vehicles and equipment.
(iii) Falls from the loading docks
   Loading dock and ramp can be considered as dangerous areas. They are frequently congested, heavy-traffic areas, and the working and walking surfaces are usually wet. Metal dock plates tend to be very slippery and the edge of the dock plates will initiate trips and falls.
(iv) Fall down the stairs
   Stairwells should be well-lighted with handrails on both sides. A person using the stairwells should have one hand free to have the capacity to utilise the handrails.

1.3.2   Same-Level Fall

In previous study [18], same-level fall is classified into three categories namely (i) slip and fall, (ii) trip and fall, and (iii) step and fall. Slip-and-fall incidents tend to occur due to the insufficient coefficient of friction (COF) or slip resistance between the shoe and the floor. A sliding motion occurs when the centre of gravity remains behind the foot slip and does not get the necessary support during the heel strike. This condition may cause a person to become unbalance and has the tendency to fall [19, 20].

(i)   Slip-type falls (Slip and fall)
      This incident occurs when there is an insufficient coefficient of friction or slip resistance between a person's foot and the floor. Usually, during the heel strike, a sliding motion will occur where the centre of gravity remains behind the foot slip and it becomes unsupported. Hence, unbalance occurs and gives a potential to fall.
(ii) Trip-type of falls (Trip and fall)
      The incident occurs when a person encounters an unnoticed raised object or surface in one's walking path. A trip-type fall may typically occur when there is any objects or raised characters on the walking surface that prevent leg from swinging forward to support the upper body.
(iii) Step-down-type fall (Step and fall)
      Step and fall incident happens when the foot encounters an unexpected step down, a hole, or a pressure on a walking surface. It occurs when one experiences an unnoticed single step down placed in a regularly foreseen level surface.

1.4   Slip-and-Fall Approaches

Suitable approaches are important to understand the mechanisms involved in slip-and-fall incident. There are four approaches that can help a person to better understand any slip-induced fall accidents, namely (i) epidemiology, (ii) biomechanics, (iii) tribology, and (iv) human-centred [21].

1.4.1   Epidemiology Approach

Epidemiology approach deals with the identification of incidents, distribution, and potential risk factors for injuries in a population [22]. The accidents determined by this approach are related to health condition [23], mental fatigue [24], age [25, 26], and safety management [27–29]. Other than that, previous studies stressed on the slip potential models which are related to several factors such as environment, human factors, footwear, flooring, cleaning, and contaminants [30]. Deep analysis of each factor is important in order to take any actions for the purpose of prevention.

In Malaysia, manufacturing industry faces the highest number of industrial accidents where manufacturing workers are highly exposed to the risk of slip-and-fall [11]. Slip-induced fall is the common cause of accidents in manufacturing industry [4]. Slip-and-fall has been studied by many researchers in various field such as at the cafeteria [31], restaurant [32, 33], and university campus [34].

1.4.2 Biomechanics Approach

The study of biomechanics is usually concerned with gait parameters such as heel velocity, shoe angle, as well as horizontal and vertical forces pertaining to heel strike [35, 36]. Previous researchers confirmed the importance of dynamic data of human to improve the devices capability of slip resistance measurement [37, 38]. Biomechanics of slip becomes an important component in preventing fall-related injuries [39]. It is a huge challenge to balance our system because humans are always moving on the ground with one foot in contact during walking, no feet in contact during running, or both feet in contact when standing [40].

In terms of biomechanical aspects of slip, most researchers in the past concerned on human responses to the unexpected contaminants occurred on the floor surface [41, 42]. Some studies are concerned with the measurements of kinematics that are related to human slips such as parameters of displacement, velocities, and body part positions [33, 43–45]. There is also a research studying the kinematics before a slip event occurs, which is associated with the measured friction coefficient [46]. Meanwhile, another previous study analysed the outcomes of slip event [47].

1.4.3 Tribological Approach

Tribological approach deals with the process of surface dissipative [48]. This approach is all about hydrodynamics of contaminants between shoes, floor, and the viscoelastic characteristics of the shoe materials. Fall-accident prevention usually focuses on the static and dynamic coefficient of friction between the shoe and the floor surface [36]. In previous study, the correlation between subjective perception of slipperiness and the level of friction was claimed to be significant [49].

Previous study also revealed that the shoe surface and the floor interface may influence the friction under the liquid-contaminated conditions [50]. Meanwhile, other researchers discovered that the roughness of surface and the waviness of parameter have strong correlation with the measured friction [51]. If there is any fluid exists between two sliding surfaces, it will provide lubrication, hence reduce the dynamic coefficient of friction values [52]. In addition, the reason why surface roughness measurement correlates with dynamic COF values is because of the deformation forces; the rougher the floor surface, the higher the COF [53].

1.4.4   Human-Centered Approach

Objective approach, subjective approach, or a combination of both approaches can be used to measure slipperiness and to estimate slipping events [54]. These can be used to explore slip-and-fall from the start of a slip event, the loss of balance, and the falling itself [55]. In addition, previous researchers also claimed that these approaches complement the mechanical friction-based test methods [55] (Fig. 1.4).

Psychophysical and organisational are there presentations of human-centred approach. Psychophysical indicates the perception of slipperiness with visual and tactile cues while organisational concerns about macro ergonomics and safety climate [56].

In order to assess slipperiness, visual cues and feedback in perceptions can be considered. When potential hazard conditions are being perceived by visual and tactile cues, the walking gait of human will automatically be adjusted [57]. The visual aspect becomes an important psychophysiological parameter in the gait regulation [56]. Previous researchers agreed that the perception of floor slipperiness can influence the gait pattern and human balance which can avoid a person from slipping and falling on a slippery surface [54, 58, 59].

It is also found that people depend on 'shine' information such as function of surface colour, brightness, and distance view to perform judgments of slipperiness [60]. Other study indicates that perceived reflectiveness is the strongest predictor of visually-based ratings of perceived slipperiness [61]. In addition, judgments based on texture and traction are the strongest predictive measurements with measured coefficient of friction.

Foot movement or postural instability can be considered to quantify the level of slipperiness. In the beginning of heel contact during walking, humans are usually not aware of the slip between the contaminated floor surface and the footwear [19].

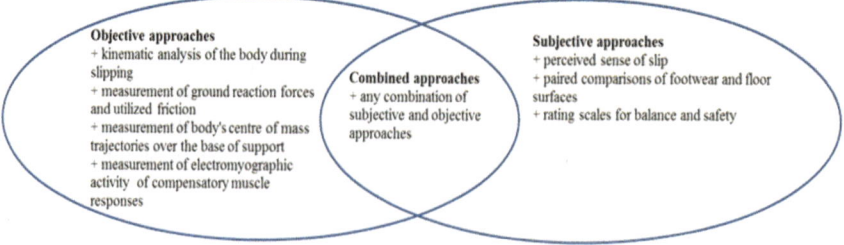

Fig. 1.4   Human-centered approach [54]

To measure the kinematic of human gait, force plates can be used. It is able to measure the slipperiness by considering ground reaction forces. Subjective ratings are also found to be correlated with the static or dynamic coefficient of friction and slip distance [49].

References

1. Balasubramaniam, A. (2012, 29 April). MTUC: Workplace accidents increasing. New Straits Times
2. Malaysia Social Security Organization (SOCSO). (2014). Malaysia Ministry of Human Resources, "Annual Repost 2014". Available from: http://www.perkeso.gov.my/
3. Malaysia Social Security Organization (SOCSO). (2013). Malaysia Ministry of Human Resources, "Annual Repost 2013". Available from: http://www.perkeso.gov.my/
4. Malaysia Social Security Organization (SOCSO). (2012). Malaysia Ministry of Human Resources, "Annual Repost 2012". Available from: http://www.perkeso.gov.my/
5. Department of Occupational Safety and Health (DOSH). (2013). Malaysia Ministry of Human Resources, "Occupational Accidents Statistics by Sector 2013". Available from: http://www.dosh.gov.my/
6. Department of Occupational Safety and Health (DOSH). (2014). Malaysia Ministry of Human Resources, "Occupational Accidents Statistics by Sector 2014". Available from: http://www.dosh.gov.my
7. Malaysia Social Security Organization (SOCSO). (2014). Malaysia Ministry of Human Resources, "Annual Report 2014". Available from: http://www.perkeso.gov.my/
8. Jaem, R. R. (2013, 8 July). Help reduce work-related accidents-riot. The Borneo Post Online.
9. Thye, L. L. (2014, 5 September). Strive for accident-free workplace environment. New Straits Times.
10. Department of Occupational and Safety (DOSH). (2016). Malaysia Ministry of Human Resources, "Occupational Accidents Statistics by Sector 2016". Available from: http://www.dosh.gov.my/
11. Department of Occupational and Safety (DOSH). (2015). Malaysia Ministry of Human Resources, "Occupational Accidents Statistics by Sector 2015". Available from: http://www.dosh.gov.my/
12. Ali S. A. S. (2010). Analysis of slips and falls among workers at workplace. Master Degree, Universiti Teknologi Malaysia, Skudai.
13. Johari, M. F. (2015). *Development of direct to indirect cost ratio of occupational accident for manufacturing industry Master Degree*. Skudai: Universiti Teknologi Malaysia.
14. Hassan, M. N., Afroz, R., Muhammad, A. F., & Awang, M. (2005). Economic instruments for managing industrial waste in Malaysia. *Malaysian Journal of Environmental Management, 6*, 87–106.
15. Economic Planning Unit. (1976). *Third Malaysia Plan 1975–1980*. Kuala Lumpur: Economic Planning Unit.
16. Ministry of Finance. (2010). *Economic Report 2010/2011*. Kuala Lumpur: Ministry of Finance.
17. Department of Occupational and Safety (DOSH). (2013). Malaysia Ministry of Human Resources, "Cause of Accident 2013". Available from: http://www.dosh.gov.my/
18. Lehtola, C. J., Becker, W. J., & Brown, C. M. (1990). *Preventing injuries from slips, trips and falls*. Institute of Food and Agriculture Science: University of Florida.
19. Perkins, P. J. (1978). Measurement of Slip Between the Shoe and Ground During Walking, In: C. Anderson, & J. Senne (Eds.), Walkway Surface of measurement of slip resistance (Vol. 649, pp. 71–87). ASTM STP.

20. Cavagna, G. A., Willems, P. A., & Heglund, N. C. (2000). The role of gravity in human walking: pendular energy exchange. *External Work and Optimal Speed., 528*(Pt 3), 657–668.
21. Lockhart, T. E. (2008). An integrated approach towards identifying age-related mechanisms of slip initiated falls. *Journal Electromyogr Kinesiol, 18*(2), 205–217.
22. Li, K. W. (1991). A biomechanical study of slipping accidents with load carriage. Ph.D. Dissertation, Texas Tech University, Lubbock, Texas.
23. Swaen, G., Burns, C. J., Collins, J. J., Bodner, K. M., Dizor, J. F., Craun, B. A., et al. (2014). Slips, trips and falls at a chemical manufacturing company. *Occupational Medicine, 64,* 120–125.
24. Lew, F. L., & Qu, X. (2014). Effects of mental fatigue on biomechanics of slips. *Ergonomics.* doi:10.1080/00140139.2014.937771.
25. Kemmlert, K., & Lundholm, L. (2001). Slips, trips and falls in different work groups-with reference to age and from a preventive perspective. *Applied Ergonomics., 32,* 149–153.
26. Moyer, B. E., Chambers, A. J., Redfern, M. S., & Cham, R. (2006). Gait parameters as predictors of slip severity in younger and older adults. *Ergonomics, 49*(4), 329343.
27. Bentley, T., Tappin, D., Moore, D., Legg, S., Ashby, L., & Parker, R. (2005). Investigating slips, trips and falls in the New Zealand dairy farming sector. *Ergonomics, 48*(8), 1008–1019.
28. Bentley, T., & Haslam, R. (2001). Identification of risk factors and countermeasures for slip, trip and fall accidents during the delivery of mail. *Applied Ergonomics., 32,* 127–134.
29. Bentley, T., & Haslam, R. (1998). Slip, trip and fall accidents occurring during the delivery of mail. *Ergonomics, 41*(12), 1859–1872.
30. Health and Safety Executive (2013). Slip potential model. Available from; http://www.hse. gov.uk/
31. Yu, R., & Li, K. W. (2013). A field assessment of floor slipperiness in a student cafeteria. *International Journal of Injury Control and Safety Promotion, 20*(3), 245–253.
32. Courtney, T. K., Huang, Y. H., Verma, S. K., Chang, W. R., Li, K. W., & Filiaggi, A. J. (2006). Factors influencing restaurant worker perception of floor slipperiness. *Journal of Occupational and Environmental Hygiene, 3,* 592–598.
33. Courtney, A. H., & Lockhart, T. E. (2012). Evaluation of gait and slip parameters for adults with intellectual disability. *Journal of Biomechanics, 45,* 2337–2341.
34. Li, K. W., Chang, W. R., Leamon, T. B., & Chen, C. J. (2004). Floor slipperiness measurement: Friction coefficient, roughness of floors, and subjective perception under spillage conditions. *Safety Science, 42,* 547–565.
35. Herman, R., Wirta, R., Bampton, S., & Finley, F. R. (1976). Human solutions for locomotion. I. Single limb analysis. In: R. Herman, S. Grillner, P. S. G Stein., & Stuart, D.G. (Eds.), Neural control of locomotion (Vol. 18). Advances in behavioral Biology. , New York: Plenum Press.
36. Lockhart, T. E. (1997). Biomechanics of slips and falls in the elderly. Master Degree, Texas Tech University.
37. Miller, J. M. (1983). Slippery work surface: Towards a performance definition and quantitative COF criteria. *Journal of Safety Research., 14,* 145–158.
38. James, D. I. (1983). Rubber and plastic in shoe and flooring: The importance of kinetic friction. *Ergonomics, 26*(1), 83–99.
39. Redfern, M. S., Cham, R., Perczak, K. G., Gronqvist, R., Hirvones, M., Lanshammar, H., et al. (2001). Biomechanics of slips. *Ergonomics, 44*(13), 1138–1166.
40. Winter, D. A. (1995). Human balance and posture control during standing and walking. *Gait and Posture, 3,* 193–214.
41. Qu, X., Hu, X., & Lew, F. L. (2012). Differences in lower extremity muscular responses between successful and failed balance recovery after slips. *International Journal of Industrial Ergonomics, 42,* 499–504.
42. Chambers, A. J., & Cham, R. (2007). Slip-related muscle activation patterns in the stance leg during walking. *Gait and Posture., 25,* 565–572.
43. Hu, X., & Qu, X. (2013). Differentiating slip-induced falls from normal walking and successful recovery after slips using kinematic measures. *Ergonomics, 56,* 856–867.

44. Callisaya, M. L., Blizzard, L., McGinley, J. L., & Srikanth, V. K. (2012). Risk of falls in older people during fast-walking-The TASCOG study. *Gait and Posture, 36*, 510–515.
45. Beschorner, K., & Cham, R. (2008). Impact of joint torques on heel acceleration at heel contact, a contributor to slips and falls. *Ergonomics, 51*(12), 1799–1813.
46. Zamora, T., Alcantara, E., Artacho, M. A., & Valero, M. (2011). Existence of an optimum dynamic coefficient of friction and the influence on human gait variability. *International Journal of Industrial Ergonomics, 41*, 410–417.
47. Cham, R., & Redfern, M. S. (2002). Changes in Gait When anticipating slippery floors. *Gait and Posture, 15*, 159–171.
48. Andres, R. O., and O'Connor, D. O. (1992). A practical synthesis of biomechanical results to prevent slip and falls in the workplace. *Advances in Industrial Ergonomics and Safety IV* 1001–1006.
49. Chang, W. R., Huang, Y. H., Li, K. W., Filiaggi, A., & Courtney, T. K. (2008). Assessing slipperiness in fast food restaurants in the USA using friction variation, friction level and perception rating. *Applied Ergonomics., 39*(3), 359–367.
50. Chang, W. R., Kim, I. J., Manning, D. P., & Bunterngchit, Y. (2001). The role of surface roughness in the measurement of slipperiness. *Ergonomics, 44*, 1200–1216.
51. Chang, W. R. (2004). A statistical model to estimate the probability of slip and fall incidents. *Safety Science, 42*(9), 779–789.
52. Chaffin, D. B., Woldstad, J. C., & Trujillo, A. (1992). Floor/shoe slip resistance measurement. *American Industrial Hygiene Association Journal, 53*, 283–289.
53. Grönqvist, R., Roine, J., Korhonen, E., & Rahikainen, A. (1990). Slip resistance versus surface roughness of deck and other underfoot surfaces in ships. *Journal of Occupational Accidents, 13*, 291–302.
54. Grönqvist, R., Abeysekera, J., Newman, D. J., Gielo-Perczak, K., Lockhart, T. T., & Pai, C. Y. C. (2001). Human-centred approaches in slipperiness measurement. *Ergonomic, 44*(13), 1167–1199.
55. Chang, W. R. (2001). The effect of surface roughness and contamination on the dynamic friction of porcelain tile. *Applied Ergonomics, 32*(1), 73–184.
56. Chang, W. R., Leclercq, S., Haslam, R., & Lochart, T. (2014). The state of science on occupational slips, trips and falls on the same level. *Industrial Health, 52*, 379380.
57. Cappellini, G., Ivanenko, Y. P., Dominici, N., Poppele, R. E., & Lacquaniti, F. (2010). Motor patterns during walking on a slippery walkway. *Journal of Neurophysiology, 103*(2), 746–760.
58. Courtney, T. K., Huang, Y. H., Verma, S. K., Chang, W. R., Li, K. W., & Filiaggi, A. J. (2006). Factors influencing restaurant worker perception of floor slipperiness. *Journal of Occupational and Environmental Hygiene, 3*, 592–598.
59. Leclercq, S. (1999). The prevent of slipping accidents: A review and discussion of work related to the methodology of measuring slip resistance. *Safety Science, 31*, 95–125.
60. Joh, A. S., Adolph, K. E., Campbell, M. R., & Eppler, M. A. (2006). Why walkers slip: Shine is not a reliable cue for slippery ground. *Perception and Psychophysics, 68*, 339–352.
61. Lesch, M. F., Chang, W. R., & Chang, C. C. (2008). Visually based perceptions of slipperiness: Underlying cues, consistency and relationship to coefficient of friction. *Ergonomics, 51*(12), 1973–1983.

Chapter 2
Factors Leading to Slip-and-Fall Incidents

Abstract There are several factors that prompt the slip-and-fall incidents and this study focuses on seven factors. Epidemiology approach was used to determine the factors of slip-and-fall incidents, and questionnaire was selected as an instrument for data collection. Exploratory Factor Analysis (EFA) was performed to evaluate how the respondents perceived the slip-and-fall factors while Cronbach's alpha was used to assess the internal consistency for each domain in the survey. Subsequently, the factors were ranked in accordance with the total variance recorded.

2.1 Seven Factors of Slip-and-Fall

In this study, the seven factors consisted of (i) footwear factor, (ii) contaminant factor, (iii) flooring factor, (iv) cleaning factor, (v) environment factor, (vi) individual factor, and (vii) perception of risk factor. Statistical analysis was important to validate how the respondent perceived the slip-and-fall factors in this study.

In the analysis, Exploratory Factor Analysis (EFA) was chosen along with Principle Components Analysis (PCA) and Varimax rotation method where Kaiser-Meyer-Olkin's (KMO) verification of sampling adequacy was performed. The eigenvalue of higher than 1.0 was set and loading item greater than 0.4 was considered. In the beginning, the EFA was conducted on 30 items of slip-and-fall factors. Then, items with low reliability and factor loading were removed. Finally, only 25 items were considered for the analysis. Based on KMO test results, the pattern of correlation was considered as accepted (0.701, mediocre). Then, the factor loading values were obtained (ranging from 0.345 to 0.930). The Bartlett's test of sphericity recorded statistical significance with $\chi^2(435) = 1325$, $p < 0.0001$, indicating that the correlation between the items was sufficient for PCA. This kind of information is also supported by previous study [1]. It was found that the seven factors solution in the analysis occupied 59.809% of the total variance.

© The Author(s) 2017
N.A. Ahmad et al., *Quantitative and Qualitative Factors that Leads to Slip and Fall Incidents*, SpringerBriefs in Applied Sciences and Technology, DOI 10.1007/978-981-10-3286-8_2

2.2 Leading Factors by Rank

Cronbach's alpha was utilised to verify the interrelatedness of items [2]. There liability analysis used it to find the internal consistency for each domain of the survey. Cronbach's alpha discovered that the values were ranging from 0.673 to 0.894 in the analysis. Hence, the seven factors of slip-and-fall incidents were considered reliable and obtained high internal consistency. This consideration was made based on previous study [1].

To describe a set of 'p' random variables (factors), factor analysis was used where interpretation of coefficient in a factor model (loading) can be determined. The factor loading in this analysis was found in a value ranging from 0.345 to 0.93.

Based on workers' perception of the seven factors, the analysis showed that the perception of risk was the main factor of slip-and-fall (12.605%), followed by environment factor (10.161%), individual factor (9.471%), cleaning factor (9.039%), contaminant factor (7.333%), footwear factor (6.025%), and flooring factor (5.176%). Based on the total variance obtained, those tested factors were ranked as follows: (1) perception of risk factor; (2) environment factor; (3) individual factor; (4) cleaning factor; (5) contaminant factor; (6) footwear factor; and (7) flooring factor. Table 2.1 presents factors that are established based on the percentage of variance.

In this study, perception of risk factor was identified as a significant factor of slip-and-fall incidents. Various researchers in previous studies considered the perception of risk to avoid slip-and-fall hazard before it occurs [3–7]. Based on the findings, the experienced workers tend to ignore their own safety at the workplace. In this case, being too confident is among the causes of accidents to happen.

As stated in the previous study, personality plays a big role in the way people react to instructions, either by ignoring or complying with them [8]. Previous researchers stated that when some people are unable to perceive risks well (poor perception of risks), they will not maintain their body balance and cannot adjust their gait pattern [5, 9, 10]. As found in this study, new workers are also prone to getting slip-and-fall incidents because they are not familiar with their own working environment. They lack the experience and understanding of their health and safety responsibilities. This finding is supported by earlier studies that highlighted the lack of experience [11] and the lack of understanding on health and safety [12]. In this condition, it is important for the management to take preventive measures. According to previous study, poor perception of risk by the workers and failure to provide health and safety training to the workers make the accidents unavoidable [12]. Thus, some researchers suggested a good safety management by managing the perception of risk that may provide a better organisational safety culture [13, 14].

Environment is the second factor that contributes to the slip-and-fall incidents in the manufacturing industry. In this study, respondents agreed that poor lighting, noisy environment, unpleasant condition, unsuitable temperature, and glare are the criteria of this factor. In a working area, glare or insufficient lighting may hamper the workers from seeing clearly [8, 15, 16]. Unsuitable temperature during working

Table 2.1 Summary of factor and reliability analyses on slip-and-fall factors

	Factors and items	Factor loading	% of variance	Cronbach's alpha, α
	Footwear factor			
1.	Wearing shoes with non-skid soles may avoid slip-and-fall incidents	0.817	6.025	0.743
2.	Wearing unsuitable shoes at the workplace must be avoided	0.864		
3.	Old shoes that are already worn out may cause a person to slip	0.751		
	Contaminant factor			
4.	Spillages or items of contamination attended at all times give hazard	0.821	7.333	0.673
5.	All spills must be wiped up promptly	0.854		
6.	Grease and oil accumulations cannot be ignored	0.717		
7.	Most spillages occur as a result of faulty machineries	0.475		
	Flooring factor			
8.	Suitable flooring must be used at the work areas	0.708	5.176	0.692
9.	Applications of polishes and coatings can change surface properties	0.569		
10.	Slippery floor may cause a person to slip	0.721		
	Cleaning factor			
11.	Wet floor warning or caution sign must be displayed	0.778	9.039	0.764
12.	Spillages cannot be attended at all times	0.696		
13	Wet floor must be fenced off until dry	0.830		
14.	Working area must be in tidy	0.729		
	Environment factor			
15.	Poor lighting contributes to the slip-and-fall incidents	0.747	10.161	0.771
16.	Noisy and unpleasant conditions may cause distractions	0.849		
17.	Unsuitable temperature during working hours can reduce alertness	0.827		
18.	Glare may cause difficulty for a person to see	0.526		
	Individual factor			
19.	Tiredness is one of the factors that initiates accidents	0.93	9.471	0.74
20.	Stress or pressure may interfere a person during working hours	0.927		
21.	Avoid carrying loads that block your vision	0.345		
22.	A person who has any health or medical problems may easily be involved in accidents	0.604		

(continued)

Table 2.1 (continued)

	Factors and items	Factor loading	% of variance	Cronbach's alpha, α
	Perception of risk factor			
23.	Too confident is one of the causes of slip-and-fall incident to occur	0.863	12.605	0.894
24.	Experienced workers often ignore their safety at the workplace	0.919		
25.	Unfamiliarity with own working condition and environment may cause you to be involved in slip-and-fall incidents	0.912		

[8] and noisy environment [8, 16, 17] interrupt workers' awareness in regard to any hazards in a non-ergonomic condition. Humidity is another contributing factor in the environment. Under a humid environment, the floors may be slippery [8]. All of these findings are supported by previous studies [15–17].

Next, the third factor is the individual factor. In this study, the respondents agreed that tiredness, poor stress management, improper working method, and poor health condition have made them prone to slipping and falling. These findings are consistent with previous study that discovered improper working methods such as carrying load can block one's own vision [18]. Poor health condition [4, 19, 20] and lost footing balance [12] were also found to be the criteria in slip-and-fall incidents. In addition, several earlier studies have confirmed that other criteria of individual personality such as age [19, 20, 22], experience [20], gender [22], and body mass [20, 23] are important in order to prevent slip-and-fall incidents. These criteria were considered as the background details in the studies. For example, two workers may have the same weight but different health conditions, or they may adopt different working methods to suit their individual personality. In conclusion, individual factor plays a big role in adopting practical strategies to avoid slip-and-fall incidents.

Cleanliness is the fourth factor of slip-and-fall incidents. The respondents of this study opined that wet floor must be fenced off and allocated with a warning sign, indicating that effective solutions for this factor are rather simple and inexpensive, and can ignite other benefits of preventing slip-and-fall incidents. This statement is supported by previous study [24]. Spillage cannot be attended at all times hence a working area must always be tidy. In this case, [24] agreed on the importance of removing spillage while other researcher emphasised the tidiness aspect of a working area [21]. These studies particularly measured the relationship between causes and injury severity. Inadequate cleaning is the cause of most injury cases that are severe and consequently require intensive treatment [21].

The fifth factor in this study is contaminant. A contaminant can be represented by various types of liquid such as detergent [17, 25], water [25], and oil [17, 18, 25]. In this study, the respondents agreed that any contamination can cause hazard hence it must be cleaned immediately. These findings corroborate the findings of previous study [6]. The respondents also agreed that grease and oil accumulation cannot be ignored because most of the spillages in manufacturing industry is a result of faulty machinery. In previous study, oil contaminant received the lowest COF value hence it should be avoided [6]. In addition, it was found that most spillages occur as a result of faulty machineries [26] that do not have adequate maintenance [21]. These contaminants are dangerous as they can initiate slip-and-fall incidents [16, 17, 25, 28, 29].

A person can slip on an oil contaminant due to the lowest value of friction being measured compared to water [25]. Hence, the accumulation of grease and oil cannot be ignored by workers in the manufacturing industry. Moreover, contamination control needs to be applied because almost all slip-and-fall incidents occur due to contamination [30]. Respondents' perception in this case is consistent with the findings in the previous studies [17, 25, 27–29].

Footwear is the sixth factor in any slip-and-fall incidents. In this study, it was found that non-skid soles may avoid slip-and-fall incidents and it was recommended that unsuitable shoes must be avoided [24, 28, 31]. This is because different types of job, floor surface, floor condition, and sole material have different purposes and functions [24]. Hazards or accidents are unavoidable if workers wear shoes that are unsuitable with their working condition and environment. In this study, worn-out shoes were also perceived as a cause of slipping, as mentioned in previous study [24]. Therefore, each person should be alert with the condition of their shoes. Previous researchers also considered sole material in preventing slipperiness [25, 26, 28]. The criteria in this factor are supported by earlier studies [24–26, 28, 31].

Finally, flooring is the seventh factor contributing to the slip-and-fall incidents. Most respondents agreed that slippery floor may cause them to slip and fall. This argument is consistent with [25]. In addition, suitable flooring in terms of floor selection must be considered at the workplace [24]. Different types of floor are needed for different working environments to ensure the safety of workers and prevent them from involving in any accidents. The applications of polish and coating can change the surface properties, where the floor finishes may help reduce the floor's slippery condition [32]. Hence, it is important to understand weather a surface can be modified to improve slip resistance. All of the findings were related to the friction and roughness of the floor, which is a factor that will affect the potential of slip-and-fall incidents. Earlier studies confirmed the various issues related to floor friction [25, 28, 33] and floor roughness [6, 32, 34, 35]. Thus, a flooring factor can be considered to study the effect of slipperiness in the slip-and-fall areas.

References

1. Field, A. (2005). *Discovering statistics using SPSS* (2nd ed.). London: SAGE Publications.
2. Tavakol, M., & Dennick, R. (2011). Making sense of cronbachs alpha. *International Journal of Medical Education, 2*, 53–55.
3. Courtney, T. K., Verma, S. K., Chang, W. R., Huang, Y. H., Lombardi, D. A., Brennan, M. J., et al. (2013). Perception of slipperiness and prospective risk of slipping at work. *Occupational Environmental Med, 70*, 35–40.

4. Bourque, L. B., Shen, H., Dean, B. B., & Kraus, J. F. (2007). Intrinsic risk factors for falls by community-based seniors: implications for prevention. *International Journal of Injury Control and Safety Promotion, 14*, 267–270.
5. Courtney, T. K., Huang, Y. H., Verma, S. K., Chang, W. R., Li, K. W., & Filiaggi, A. J. (2006). Factors influencing restaurant worker perception of floor slipperiness. *Journal of Occupational and Environmental Hygiene, 3*, 592–598.
6. Li, K. W., Chang, W. R., Leamon, T. B., & Chen, C. J. (2004). Floor slipperiness measurement: Friction coefficient, roughness of floors, and subjective perception under spillage conditions. *Safety Science, 42*, 547–565.
7. Haslam, R. A., & Bentley, T. A. (1999). Follow-up investigations of slip, trip and fall accidents among postal delivery workers. *Safety Science, 32*, 33–47.
8. Health and Safety Executive. (2009). Slips and trips e-learning package. Available on: http://www.hse.gov.uk/slips/Step/
9. Grönqvist, R., Abeysekera, J., Newman, D. J., Gielo-Perczak, K., Lockhart, T. T., & Pai, C. Y. C. (2001). Human-centred approaches in slipperiness measurement. *Ergonomic., 44*(13), 1167–1199.
10. Leclercq, S. (1999). The prevent of slipping accidents: A review and discussion of work related to the methodology of measuring slip resistance. *Safety Science, 31*, 95–125.
11. Kong, P. W., Suyama, J., & Hostler, D. (2013). A review of risk factors of accidental slips, trips, and falls among firefighters. *Safety Science, 60*, 203–209.
12. Peebles, L., Wearing, S., & Heasman, T. (2005). *Identifying Human factors associated with slip and trip accidents* (1st ed.). London: Health and Safety Executive.
13. Cooper, M. D. (2000). Towards a model of safety culture. *Safety Science, 36*, 111–136.
14. Gill, G. K., & Shergill, G. S. (2004). Perceptions of safety management and safety culture in the aviation industry in New Zealand. *Journal of Air Transport Management, 10*, 233–239.
15. Lehtola, C. J., Becker, W. J., & Brown, C. M. (1990). Preventing injuries from slips, trips and falls. Institute of Food and Agriculture Science, University of Florida.
16. Hu, X., & Qu, X. (2013). Differentiating slip-induced falls from normal walking and successful recovery after slips using kinematic measures. *Ergonomics, 56*, 856–867.
17. Cham, R., & Redfern, M. S. (2002). Changes in gait when anticipating slippery floors. *Gait and Posture, 15*, 159–171.
18. Myung, R., & Smith, J. L. (1997). The effect of load carrying and floor contaminants on slip and fall parameters. *Ergonomics, 40*(2), 235–246.
19. Chau, N., Gauchard, G. C., Siegfried, C., Benamghar, L., Dangelzer, J. L., Francais, M., et al. (2004). Relationships of job, age and life conditions with the causes and severity of occupational injuries in construction workers. *International Archives of Occupational and Environmental Health, 77*, 60–66.
20. Kong, P. W., Suyama, J., & Hostler, D. (2013). A review of risk factors of accidental slips, trips, and falls among firefighters. *Safety Science, 60*, 203–209.
21. Peebles, L., Wearing, S., & Heasman, T. (2005). *Identifying human factors associated with slip and trip accidents* (1st ed.). London: Health and Safety Executive.
22. Kemmlert, K., & Lundholm, L. (2001). Slips, trips and falls in different work groups-with reference to age and from a preventive perspective. *Applied Ergonomics, 32*, 149–153.
23. Froom, P., Melamed, S., Kristal-Boneh, E., Gofer, D., & Ribak, J. (1996). Industrial accidents are related to relative body weight: the Israeli CORDIS study. *Occupational and Environmental Medicine, 53*, 832–835.
24. Saif. (2008). Prevent slips, trips and fall—if you noticed a hazard, ACT. Saif Corporation, United States.
25. Li, K. W., Chang, W. R., Leamon, T. B., & Chen, C. J. (2004). Floor slipperiness measurement: friction coefficient, roughness of floors, and subjective perception under spillage conditions. *Safety Science, 42*, 547–565.
26. Manning, D. P., & Jones, C. (2001). The effect of roughness, floor polish, water, oil and ice on underfoot friction: Current safety footwear solings are less slip resistant that microcellular polyurethane. *Applied Ergonomics., 32*, 185–196.

27. Hu, X., & Qu, X. (2013). Differentiating slip-induced falls from normal walking and successful recovery after slips using kinematic measures. *Ergonomics, 56*, 856–867.
28. Liu, L., Li, K. W., Lee, Y. H., Chen, C. C., & Chen, C. Y. (2010). Friction measurements on 'anti-slip' floors under shoe sole, contamination and inclination conditions. *Safety Science, 48*, 1321–1326.
29. Moyer, B. E., Chambers, A. J., Redfern, M. S., & Cham, R. (2006). Gait parameters as predictors of slip severity in younger and older adults. *Ergonomics, 49*(4), 329–343.
30. Health and Safety Executive. (2013). Slip potential model. Available from: http://www.hse.gov.uk/
31. Gao, C., Abeysekera, J., Hirvonen, M., & Gronqvist, R. (2004). Technical note—slip resistant properties of footwear on ice. *Ergonomics, 47*(6), 710–716.
32. Morrey, M. L. (2006). A study of the slip characteristics of applied epoxy resin flooring and thin coat epoxy base materials. Research Report 497: Health and Safety Executive (HSE).
33. Li, K. W., Hsu, Y. W., Chang, W. R., & Lin, C. H. (2007). Friction measurements on three commonly used floors on a college campus under dry, wet, and sand-covered conditions. *Safety Science, 45*, 980–992.
34. Chang, W. R. (2001). The effect of surface roughness and contamination on the dynamic friction of porcelain tile. *Applied Ergonomics, 32*(1), 73–184.
35. Kim, I. J., Hsiao, H., & Simeonov, P. (2013). Functional levels of floor surface roughness for the prevention of slips and falls: Clean-and-dry and soapsuds-covered wet surfaces. *Applied Ergonomics, 44*, 58–64.

Chapter 3
The Relationship Between Coefficient of Friction (COF) with Floor Slipperiness and Roughness

Abstract The relationship between coefficient of friction (COF) and roughness is best found using tribology approach. This study used experimental design to collect and measure the data on COF and the roughness measurement under contaminated condition. It particularly measured the COF of four different floor surfaces in five surface conditions—one dry condition and four liquid-spillage conditions. For slipperiness measurement, the analysis of variance (ANOVA) was used to determine the effect of floor, footwear, and surface condition on the measured COF. Duncan's multiple range tests was performed to determine the sample of means that is significantly different from others. Meanwhile, Pearson's correlation coefficient was used to measure the strength between COF and the roughness parameter.

3.1 Measuring Device

According to Health and Safety Executive (2012), a portable skid resistance tester is also known as (1) a pendulum COF test, (2) a British pendulum, or (3) a Transport and Road Research Laboratory (TRRL) pendulum. The mechanism of this instrument is based on a swinging-and-imitation heel concept. On the vertical column of the tester is a rack and pinion which are important for controlling the movement of the tester's head. The rack and pinion can also carry a swinging arm and a graduated scale. A pointer on the rear of the vertical column of the tester is important to control the movement of the tester's head that carries the swinging arm and the graduated scale through a pointer-and-release mechanism.

For the data collection, a mean of five successive readings were recorded if the data did not differ by more than three units. In the case where the difference was greater than three units, the swings were repeated until three successive readings were constant and the data could be recorded [1]. Figure 3.1 displays the portable skid resistance tester used in this study.

Floor surface texture can be used as an indicator of slipperiness in water contaminated conditions. Surface roughness can also be used to monitor the changes of

N.A. Ahmad et al., *Quantitative and Qualitative Factors that Leads to Slip and Fall Incidents*, SpringerBriefs in Applied Sciences and Technology,
DOI 10.1007/978-981-10-3286-8_3

Fig. 3.1 Munro Stanley
portable skid resistance
tester [2]

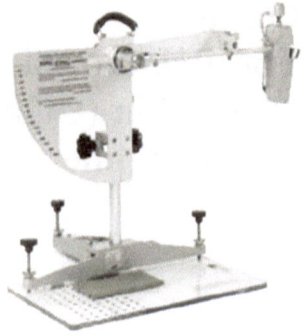

Fig. 3.2 Surftest SJ-210 [3]

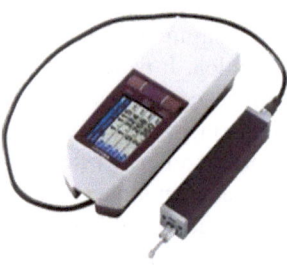

the floor surface characteristics such as wear. Figure 3.2 presents the portable
surface roughness tester used in this study.

3.2   Test Condition

This study considered four types of floor namely (i) ceramic I floor—glazed ceramic
tile (Fig. 3.3), (ii) ceramic II floor—unglazed ceramic tile (Fig. 3.4), (iii) epoxy
floor (Fig. 3.5), and (iv) porcelain floor—homogenous tile (Fig. 3.6). For each type
of floor, one area was selected for measurement. There was no standard for selecting

Fig. 3.3 Ceramic I floor

Fig. 3.4 Ceramic II floor

Fig. 3.5 Epoxy floor

Fig. 3.6 Porcelain floor

the location for friction measurement. In this study, the middle of the floor sample area was measured.

This study used three types of footwear materials namely (i) nylon (Fig. 3.7), (ii) rubber (Fig. 3.8), and (iii) polyvinyl chloride or PVC (Fig. 3.9). Samples of these materials were supplied by a shoe manufacturer. All materials used in this study were flat, as there was no tread on the samples during the testing. Figure 3.7 until Fig. 3.9 represent all the materials used in this study.

For COF measurement, using a flat footwear material may not represent the actual friction on the floor, but it would at least estimate the performance or the COF effect of the different materials. Earlier study found that a tread design has no advantage in improving the slip resistance of wet and glycerol-contaminated

Fig. 3.7 Nylon

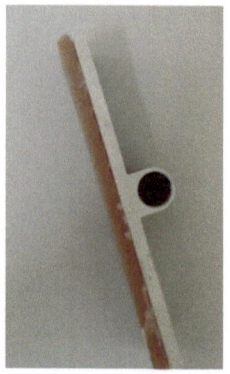

Fig. 3.8 Rubber

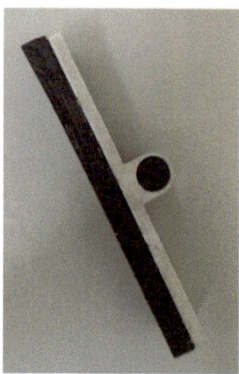

Fig. 3.9 PVC

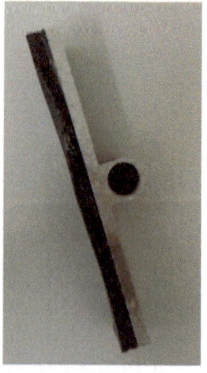

conditions [4]. This is because the liquid flows into the valleys and concaves on the floor immediately when the water is applied. Hence, the peaks on the floor only have a very thin film of water where the lubrication effects of water on the floor were also diminished [4]. That is why tread footwear gives no advantage under such a tested condition compared to flat footwear material.

Table 3.1 Viscosity value of liquid

	Water	Water and detergent	Cooking oil	Engine oil
1	0.53	0.64	65.13	328.19
2	0.96	1.01	64.58	327.99
3	1.28	1.27	64.34	328.15
4	1.67	1.6	64.14	328.21
5	1.88	1.8		328.2
6		2.53		328.16
7		2.21		327.14
Average	1.264	1.58		328.006

For the experiment, 10 ml of liquid was used to create a wet, spillage condition. As for the water contaminant, only water was used, and for the wet detergent contaminant, 5% detergent and 95% water were used. For cooking oil and engine oil contaminants, 10 ml of cooking oil and the same amount of engine oil were spread evenly on the floor. The viscosities for all types of liquid were then measured. Table 3.1 presents the viscosity value of each liquid. The result shows that engine oil appeared to be the most viscous liquid (328.006 cSt), followed by cooking oil (64.548 cSt), water and detergent (1.58 cSt), and water (1.264 cSt).

3.3   Friction and Roughness Measurement

In friction measurement, the head of pendulum was fitted with three different types of footwear material slider. These sliders were moved by a spring system releasing a pendulum out in a horizontal position. After this process, the friction values could be directly read from the measuring scale. The measurements were taken in the direction of the walking path on the selected floors (Fig. 3.10). Five measurements were taken with different footwear material/floor/testing conditions. Subsequently, a total of 300 (5 × 4 × 3 × 5) measurements were made.

Before each measurement was taken, all of the floors and footwear materials were cleaned with 50% ethanol solution. Then, the footwear materials were wiped using absorbent papers and were cleaned using 5% detergent solutions to remove the excessive contaminants. The surface was then rinsed with water and dried using a hair dryer.

Surface roughness measurement is important to determine the primary texture of surface that leads to the quantification of floor topography [5]. In this study, the surface parameters were measured using Mitutoyo Surftest 210 profilometer for the five floor surfaces. The parameters consisted of (i) average roughness (Ra), (ii) root mean square of surface heights (Rq), (iii) average peak to valley (R), and (iv) maximum peak to valley height (Rt). The measurements for each floor surfaces were taken at the four corners and in the same direction as COF measurements

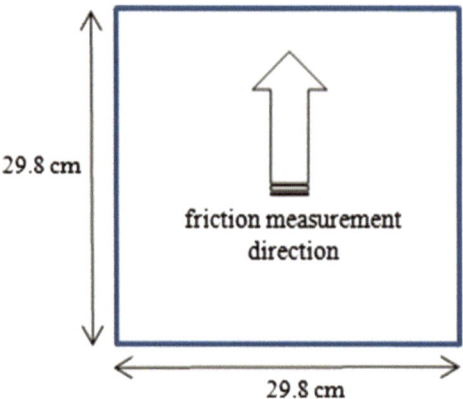

Fig. 3.10 Direction of friction measurement

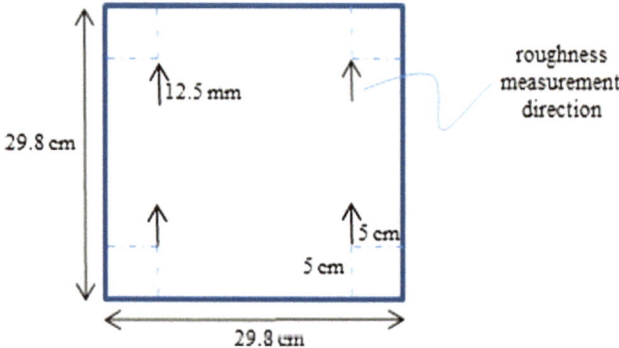

Fig. 3.11 Location and direction of roughness measurement

(Fig. 3.11). Based on the profilometer measured, the travel distance was 12.5 mm with a cut-off length of 2.5 mm. In this study, eight measurements (4 corners × 2 replications) were taken for each floor surface. The measurements taken were sufficient as previous study also adopted the same number of measurements for their roughness measurement [5]. The means and standard deviations were also calculated for the floor surfaces.

3.4 Floor Slipperiness Measurement

Analysis of variance (ANOVA) is a method used to analyse the impact of one or more nominal variables (independent variables) on a quantitative variable (dependent variable) [6]. By performing ANOVA in this study, the effect of floor, footwear, and surface condition on the measured COF can be determined. The analysis showed that the effects of the three factors on COF were statistically significant ($p < 0.0001$). All two-way and three-way interaction effects were also significant ($p < 0.0001$). This illustrates that COF depends not only on the type of floor, but also on the types of footwear and surface condition.

Generally, Duncan's multiple range tests are used after ANOVA in order to determine which sample of means that are significantly different from the other [7]. Tables 3.2, 3.3, and 3.4 present the results for floors, footwear materials, and surfaces respectively, where different letters in the groups indicate that they are significantly different under a = 0.05. In Table 3.2, it is shown that the COF values for all floors are statistically different. The ceramic I floor recorded the lowest mean of COF, followed by ceramic II floor, epoxy floor, and porcelain floor respectively. As shown in the Table 3.3, the COF values for all footwear materials are also significantly different from one another. The orders from low to high are PVC, nylon, and rubber. As for the different surface conditions (Table 3.4), floor with engine oil shows the lowest COF, followed by cooking oil, water detergent, wet, and dry floors. Constructing the means within a table (Tables 3.2, 3.3, and 3.4) gives more confidence when the means were compared. Thus, different letters in each Tables 3.2, 3.3, and 3.4 show that each treatment is significantly different from one another.

Figures 3.12, 3.13, and 3.14 represent the two-way interaction of the three factors. As shown in Fig. 3.12, the COF values were generally reduced when the

Table 3.2 Duncan's multiple range test results for floor [12]

Floor	Mean COF	Group
Ceramic I	28.48	A
Ceramic II	29.53	B
Epoxy	34.04	C
Porcelain	38.41	D

Table 3.3 Duncan's multiple range test results for footwear [12]

Footwear	Mean COF	Group
PVC	21.45	A
Nylon	32.02	B
Rubber	44.38	C

Table 3.4 Duncan's multiple range test results for surface condition [12]

Surface condition	Mean COF	Group
Engine oil	21.92	A
Cooking oil	24.03	B
Water-detergent	28.22	C
Wet	32.8	D
Dry	56.12	E

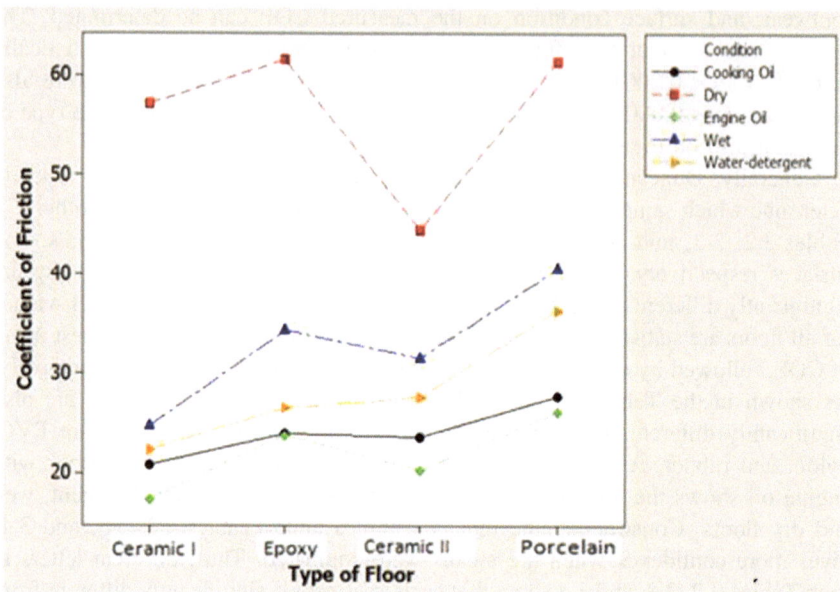

Fig. 3.12 The interaction between floors and contaminants (average COF values for each footwear material) [12]

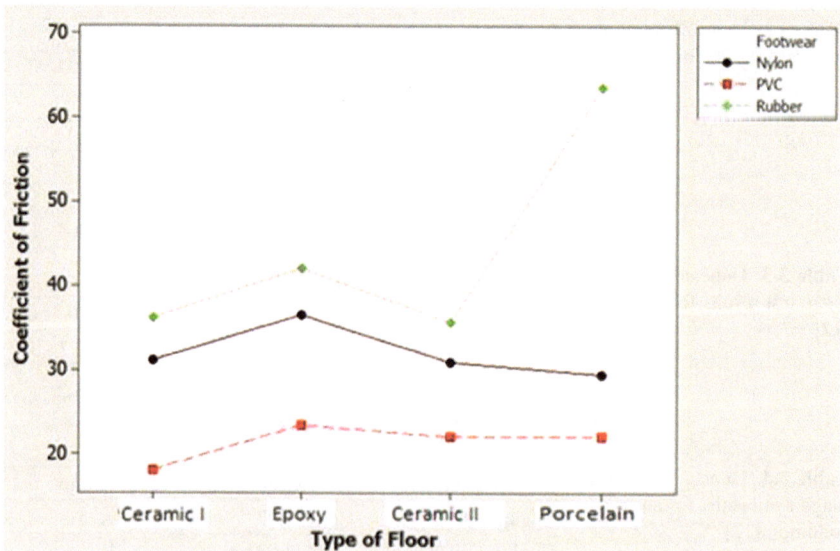

Fig. 3.13 The interaction between floors and footwear materials (average COF values for each contaminated condition) [12]

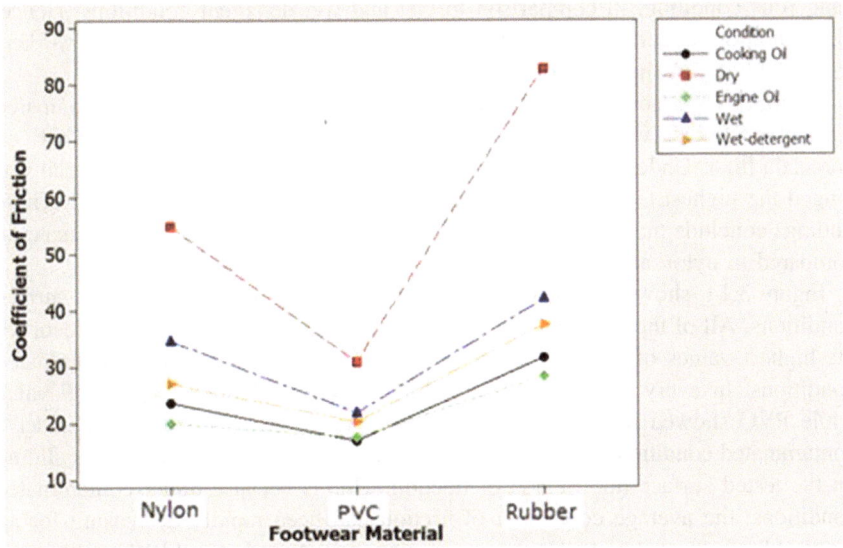

Fig. 3.14 The interaction between footwear materials and surface conditions (average COF values for each floor type) [12]

floors were covered by the liquid of any contaminants. The COF values reduced sharply on ceramic I and epoxy floors. In addition, the porcelain floor showed higher COF values in the range of 26–61 compared with other floors under liquid (wet and water detergent) and oil conditions. Therefore, it can be concluded that all tested floors had recorded the highest average coefficient of friction on dry condition (the floors were not slippery). In a contaminated condition, all tested floors became slippery. However, engine oil gave the most slippery condition, showing the lowest average coefficient of friction on each tested floor.

In this study, porcelain floor (homogenous tile), which is widely used by manufacturing firms, has better slip-resistant surface due to its very high COF values (Table 3.2) compared to ceramic I (glazed ceramic tile), ceramic II (unglazed ceramic tile), and epoxy floor. The selection of proper floor materials (due to the degree of roughness and/or geometric design of the floor surface) may be an alternative for slip-prevention purpose [8]. In this sense, due to the measured COF value, a porcelain floor (homogenous tile) is considered the best floor in manufacturing industry and a better option than ceramic I (glazed ceramic tile), ceramic II (unglazed ceramic tile), and epoxy floors.

Under all contaminated conditions, the squeeze-film effect on the COF is significant when there is an interaction between a flat footwear and a smooth floor surface [9]. This concept clearly explains the reasons of low COF values that were obtained in this study for ceramic I, ceramic II, and epoxy floors under all liquid-contaminated conditions. It can be argued that the measured COF are affected by the surface condition and in this sense, oily condition is the most dangerous condition, in comparison to wet and wet detergent conditions. Hence, oily condition in manufacturing operations must be avoided to prevent the workers from involving in slip-and-fall incidents.

In Fig. 3.13, nylon and PVC showed a higher COF on epoxy floor compared with other floors, while rubber showed a notable different value of COF on porcelain floor. Under all surface conditions, the mean COF of rubber material was ranged the highest (36–63), compared to nylon (29–37) and PVC (18–24). These findings conclude that on each tested floor, rubber is a better slip-resistant material compared to nylon and PVC.

Figure 3.14 shows the interaction between footwear materials and surface conditions. All of the footwear materials namely nylon, rubber, and PVC recorded the highest values of COF under dry condition compared to under contaminated conditions. In every contaminated condition, rubber had the highest COF value while PVC showed the lowest value. Hence, it can be understood that under a contaminated condition, all types of footwear materials tend to have a slight sliding on the tested surface due to loss of traction. This is because under contaminated conditions, the average coefficient of friction had been rapidly decreasing for all types of footwear materials. However, in comparison to nylon and PVC, rubber can be considered as the best material for all contaminated conditions. Under room temperature, rubber is the best choice of footwear material for safety footwear that is suitable to be used in the industry due to the highest COF value recorded.

The interaction graph in this study has clearly established the relationship between COF and floor slipperiness.

3.5 Floor Roughness Versus COF

In this study, the correlation coefficient is numbered from −1 to 1, indicating the strength of the linear relationship between the type of floors and the roughness parameter. In this sense, Pearson's correlation coefficient can be used to measure the strength of a linear relationship where a sign of correlation specifies the direction of the relationship [10]. Table 4.8 indicates the correlation between four roughness parameters (R_a, R_q, R_z, and R_t) and COF under different surface conditions. According to the results, the correlation between COF and roughness parameter under the conditions of wet, water detergent, and cooking oil were considered very high ($r = 0.913$–0.995) while the correlation under dry condition was very low ($r = 0.084$–0.316). In addition, under the condition of engine oil, the correlation can be considered high ($r = 0.861$–0.914).

In the previous study, a high correlation between surface parameters and slip resistance was found [11]. This statement supports the current finding where the correlation coefficient showed that the COF was well represented by floor roughness under contaminated conditions (Table 3.5). In an earlier study [5], the floor roughness did not appear as a suitable indicator of floor slipperiness under an oily condition due to the negative correlation. However, in this study, a positive

Table 3.5 Pearson's correlation coefficients (r) between floor roughness and COF under different contaminated conditions

	R_a	R_q	R_z	R_t
Dry	0.11	0.084	0.316	0.178
Wet	0.953*	0.970*	0.978*	0.990*
Water-detergent	0.913	0.918	0.992**	0.979*
Cooking oil	0.976*	0.983*	0.995**	0.994**
Engine oil	0.914	0.898	0.861	0.861
Overall	0.114	0.11	0.196	0.124

*$p < 0.005$, **$p < 0.01$

correlation was obtained for each surface condition (Table 4.8). This implies a strong relationship between roughness and the measured COF on the tested floors. Thus, roughness is considered suitable as an indicator of floor slipperiness.

References

1. Road Research Laboratory. (1969). *Road note: Instructions for using the portable skid-resistance tester* (2nd ed.). Crowthome, Berkshire: Ministry of Transport.
2. Munro Instruments. Available from: http://www.munroinstruments.com
3. Mitutoyo America. Available from: http://www.mitutoyo.com/
4. Liu, L., Li, K. W., Lee, Y. H., Chen, C. C., & Chen, C. Y. (2010). Friction measurements on 'anti-slip' floors under shoe sole, contamination and inclination conditions. *Safety Science, 48*, 1321–1326.
5. Li, K. W., Chang, W. R., Leamon, T. B., & Chen, C. J. (2004). Floor slipperiness measurement: Friction coefficient, roughness of floors, and subjective perception under spillage conditions. *Safety Science, 42*, 547–565.
6. Michael, S., Beck, L., Bryman, A., & Liao, T. F. (2004). *The sage encyclopedia of social science research methods*. Thousand Oaks: Sage Publications Inc.
7. Vogt, W. P. (2005). *Dictionary of statistics & methodology*. Thousand Oaks: Sage Publications Inc.
8. Li, K. W., Hsu, Y. W., Chang, W. R., & Lin, C. H. (2007). Friction measurements on three commonly used floors on a college campus under dry, wet, and sand-covered conditions. *Safety Science, 45*, 980–992.
9. Moore, D. F. (1972). The friction and lubrication of elastomers. In G. V. Vaynor (Ed.), *International series of monographs on materials science and technology*. Pergamon Press: Oxford.
10. Bagiella, E. (2008). *Encyclopedia of epidemiology*. Thousand Oaks: Sage Publications Inc.
11. Chang, W. R. (1999). The effect of surface roughness on the measurements of slip resistance. *International Journal of Industrial Ergonomics, 24*, 299–313.
12. Ahmad, N. A., Md Tap, M., Syahrom, A., Rohani, J. M., & Johari, M. F. (2015). Floor slipperiness measurement under spillage condition. *Jurnal Teknologi, 77*(27), 59–63.

Chapter 4
Human Perception of Slipperiness Through Measured COF

Abstract In this study, psychophysics (the subsets of human-centred approach) which includes the perception of slipperiness with visual and tactile cues (Chang et al., Ind Health 52:379380, 2014 [1]) are important to validate human perception of slipperiness. Subject rating was used as an instrument for data collection in human-centred approach. This study measured the perception of four different floor surfaces in five surface conditions—one dry condition and four liquid-spillage conditions. The concerned tested floor surfaces were (i) ceramic I (glazed ceramic tile), (ii) ceramic II floor (unglazed ceramic tile), (iii) epoxy floor, and (iv) porcelain floor (homogenous tile). Chi-square test was used to test the subjective scores of the floor slipperiness while Spearman's rank correlation coefficient was used to measure the strength of association between the subjective scores of floor slipperiness and the measured COF.

4.1 Method of Analysis

In this study, psychophysics is the perception of slipperiness with visual and tactile cues in order to validate the experimental design. Subject rating was used as an instrument for the data collection and to assess the role of visual and tactile cues in perceived slipperiness. This study involved 24 subjects, comprising of 12 males and 12 females. Each subject was brought to the location of each floor surface, where they were asked to inspect the area visually. The subjects were allowed to touch the tested floor/surface condition and determine the slipperiness on scale of 1–5; each rank respectively corresponding to "extremely slippery", "very slippery", "slippery", "somewhat slippery", and "not slippery." The evaluation must then be repeated by the subjects on the next two days for the researcher to obtain the test-retest reliability of the perception of floor slipperiness.

All 24 subjects completed questionnaires distributed by the researcher in this study. Repeating the evaluation was needed to gain consistency from the respondents in regard to the rating of slipperiness. The data obtained were analysed to identify the subject score for slipperiness.

© The Author(s) 2017
N.A. Ahmad et al., *Quantitative and Qualitative Factors that Leads to Slip and Fall Incidents*, SpringerBriefs in Applied Sciences and Technology, DOI 10.1007/978-981-10-3286-8_4

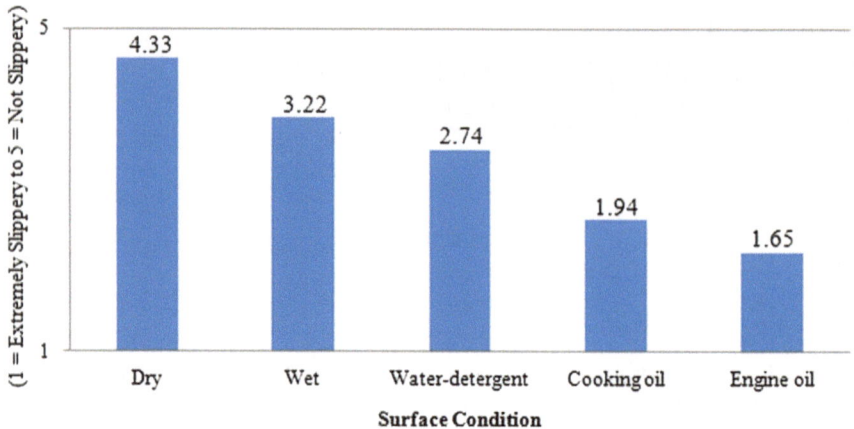

Fig. 4.1 The subjective scores of floor slipperiness under different surface conditions

4.2 Subjective Score Under Different Surface Conditions

Chi-square test is a non-parametric test of the statistical significance of a relation [2] that can be used to determine if groups in a sample are significantly different. Hence, in this study, Chi-square test was necessary to test the subjective scores of floor slipperiness. The findings revealed that the difference between the test and retest scoring was not statistically significant. Figure 4.1 shows the following mean scores and ratings for different floor conditions: (i) "not slippery" dry condition (4.33), (ii) "slippery" wet condition (3.22), (iii) "very slippery" water detergent condition (2.74), (iv)"extremely slippery" cooking oil condition (1.94), and (v) "extremely slippery" engine oil condition (1.65).

These findings totally support the results in Table 3.4 in Chap. 3 where the engine oil condition recorded the lowest friction value (21.92), followed by cooking oil condition (24.03), water detergent condition (28.22), wet condition (32.8), and dry condition (56.12). The difference between each surface condition was also significant. It is concluded that under different surface conditions, the respondents could differentiate floor slipperiness very well. The inconsistency between the types of floor and human perceptions of slipperiness implies that human subject is incapable of differentiating slipperiness, making themselves prone to slip-and-fall incidents.

4.3 Subjective Score on Different Types of Floor

Figure 4.2 shows the effect of floor type on the mean scores of floor slipperiness. The mean scores for ceramic I, ceramic II, epoxy, and porcelain floors were 2.34, 3.05, 2.46, and 3.25, implying that the respondents rated porcelain and ceramic II

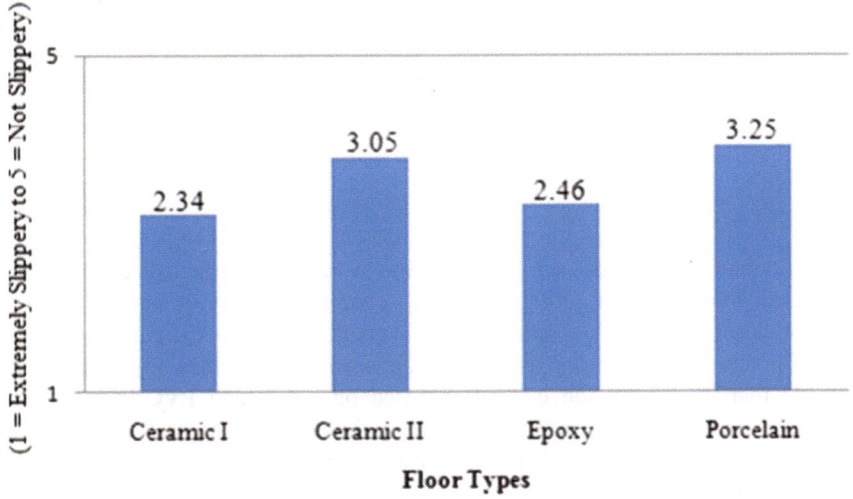

Fig. 4.2 The subjective scores of floor slipperiness on different types of floor

floors as "slippery floors" (3.25 and 3.05), while epoxy and ceramic II floors were considered "very slippery" (2.46 and 2.34).

For the type of floor, it was found that the subjective scores of floor slipperiness were not significantly different (Fig. 4.2). However, by using a skid resistance tester to compare the score with COF value (Table 3.2), the differences for each floor were found to be statistically significant. In Fig. 4.2, the mean subjective score of ceramic I was the lowest (2.34) among all tested floors, while porcelain floor recorded the highest mean score (3.25). This implies that the subjects felt that ceramic I floor was the "most slippery" and porcelain floor was the "least slippery". These results support the measured COF in Table 3.2 that records ceramic I had the lowest friction value (28.48) while porcelain had the highest friction value (38.41). On the other hand, the subjects rated epoxy floor (2.46) as more slippery than ceramic II (3.05), but in comparison to the measured COF, ceramic II floor (29.53) appeared to be more slippery than epoxy floor (34.04).

4.4   Correlation Between Perceived Slipperiness and COF

Table 4.1 presents the Spearman's rank correlation coefficient (ρ) between subjective scores of floor slipperiness and COF values measured with a portable skid resistance tester. This is the most common method used to measure the direction and strength of association between two variables [3]. In this calculation, the

Table 4.1 The Spearman's correlation coefficients (ρ) between subjective scores of floor slipperiness and COF for footwear materials and floor types

	Ceramic I	Ceramic II	Epoxy	Porcelain
Nylon	0.400	0.800	−0.632	0.200
Rubber	0.632	−0.400	0.800	−0.200
PVC	0.400	0.800	−0.600	0.600

average value of COF for each condition was used because the five friction measurements were recorded during each tested condition. Hence, it can be indicated that the closer a Spearman's rank correlation coefficient to the extremes (either 1.0 or -1.0), the stronger is the association between the variables [3]. The results showed that the correlation coefficient based on nylon and PVC materials for porcelain floor and rubber materials for epoxy floor was the highest ($\rho = 0.8$). For ceramic I, the COF-measured using rubber is the second ($\rho = 0.632$). However, the correlation coefficient based on the rubber sample on porcelain floor was very low or having no association ($\rho = 0.2$) compared to others.

The increasing negative correlation values indicate a negative relationship between the pairs [3]. As in Table 4.1, the negative correlations were found for ceramic II and porcelain floors based on rubber materials (−0.4 and −0.2 respectively) and for epoxy floor based on nylon and PVC samples (−0.632 and −0.6 respectively).The negative correlation for epoxy floor based on nylon sample was high (−0.632) compared to others. This indicates that an increasingly negative relationship existed when x variables (footwear materials) increased in value, or when y variables (floor types) decreased or remained the same. In this case, the data pairs exhibited large differences with each other. In contrast, increasingly positive relationships were noted as x variables increased in value, and y variables also increased or remained the same. This shows that the data pairs had a huge difference with one another.

According to the increasing positive correlations values, PVC is considered as the material that COF correlates best in regard to the subjective scores of floor slipperiness.

In this study, subjective ratings and measured COF complemented each other. This method is similar to the method used by previous researcher [4] who used visual-based perceptions of slipperiness to COF. It was particularly found that humans are not able to differentiate the slipperiness of different floor materials rather than under different surface conditions [4]. The findings are based on the comparison between subjective scores (Figs. 4.1 and 4.2) and COF data (Tables 3.2 and 3.4). These are consistent with those found in previous studies [5, 6]. However, these studies were lacking i n consistency in terms of the ranking of subjective scores for both floor and surface conditions.

References

1. Chang, W. R., Leclercq, S., Haslam, R., & Lochart, T. (2014). The state of science on occupational slips, trips and falls on the same level. *Industrial Health, 52*, 379380.
2. Connor-Linton, J. (2010). *Encyclopedia of research design*. Thousand Oaks: Sage Publications Inc.
3. Salkind, N. J. (2010). *Encyclopedia of research design*. Thousand Oaks: Sage Publications Inc.
4. Lesch, M. F., Chang, W. R., & Chang, C. C. (2008). Visually based perceptions of slipperiness: Underlying cues, consistency and relationship to coefficient of friction. *Ergonomics, 51*(12), 1973–1983.
5. Li, K. W., Chang, W. R., Leamon, T. B., & Chen, C. J. (2004). Floor slipperiness measurement: Friction coefficient, roughness of floors, and subjective perception under spillage conditions. *Safety Science, 42*, 547–565.
6. Cohen, H. H., & Cohen, D. M. (1994). Psychophysical assessment of the perceived slipperiness of floor tile surfaces in a laboratory setting. *Journal of Safety Research, 25*(1), 19–26.

Chapter 5
Conclusion

Abstract Achievement of research objectives are very important. In this study, the perception of risk factor was found as the main factor affecting slip-and-fall incidents. It was evident that friction was significantly affected by footwear material, type of floor, and the presence of contaminants on the floor. In addition, roughness also found consistent with the coefficient of friction (COF). According to human perception issue, the subjects could differentiate floor slipperiness under contaminated conditions but became unsure when in rating floor slipperiness for different types of floor.

This study has established the leading factors of slip-and-fall incidents. The main objective of this study was to establish the leading factors of slip-and-fall incidents. In this sense, perception of risk factor was identified as the main factor affecting such incidents. Other factors include, in an order of significance, environment factor, individual factor, cleaning factor, contaminant factor, footwear factor, and flooring factor. These findings imply that workers' perception of risk should be emphasised as a strategy to prevent slip-and-fall incidents. In addition, this study is the first to rank slip-and-fall factors in an order of significance.

It was evident that friction was significantly affected by footwear materials, types of floor, and the presence of contaminants on the floor surface. During the interactions, different types of footwear materials and floors gave different COF values, implying that suitable shoes or footwear and types of flooring are important to prevent slipping. A squeeze film effect on COF reduces a COF significantly when spill exists on any floor surfaces. Roughness was found to be consistent with the COF of the tested floors, where the correlations under contaminant conditions were considered high. Under contaminated conditions, the surface roughness parameters used in this study can serve as an alternative to measure friction using a skid resistance tester. In particular, surface roughness parameters can also rate the slipperiness of a floor. Previous studies used various types of footwear materials for testing such as leather, neolite, ethylene vinyl acetate (EVA), and rubber. This study considered other types of footwear materials such as polyvinyl chloride (PVC) and nylon.

© The Author(s) 2017
N.A. Ahmad et al., *Quantitative and Qualitative Factors that Leads to Slip and Fall Incidents*, SpringerBriefs in Applied Sciences and Technology, DOI 10.1007/978-981-10-3286-8_5

This study was also aiming to establish the human perception of slipperiness through the measured COF. Previously, only one study has so far intended to fulfil a similar aim and achieve similar results. It focused on students at one university campus as the subjects, while the present study specifically targeted industrial workers and environment. In this sense, the subjective scoring of floor slipperiness was well defined, and that the subjects could differentiate floor slipperiness under various contaminated conditions. However, they were mostly unsure when giving rating of the floor slipperiness for different types of floor.

5.1   Significance to the Researchers or Academician

The findings of this study may contribute to the industry by facilitating floor manufacturers and designers in improvising floor design or "floor reflectiveness" that can give the right perceptions. As concluded from the study, humans are not able to predict slipperiness on different floors compared to under different contaminated conditions.

5.2   Significance to the Industrial Practitioners

The findings may also help employers to develop policies and regulations on the selection of safe floor and footwear material. As found from the study, the COF interaction between nylon and PVC materials on porcelain floor (homogenous floor) and epoxy floor were low compared to rubber material (common material for safety shoes) on porcelain and epoxy floors under contaminated conditions.

5.3   Recommendations for Future Work

A biomechanics approach was not considered in this study due to unavailable equipment. The use of this approach would allow a more detailed study regarding the mechanics of fall. It may show how a subject reacts to specific physical conditions and perceptions. Individual factors that were not included in this study can be considered in future studies.

The experiment can be expanded by considering more types of floors and footwear materials used in various industries especially manufacturing. Different types of floors and footwear materials may result in different views on the interactions. The new discoveries can contribute to the body of knowledge of slip-and-fall incidents.

In this study, only four surface roughness parameters were considered. More surface parameters can be included in future studies to measure the correlation between surface parameter and different contaminant conditions.

Finally, the scope of this study can be extended to other areas or sectors such as transport, storage, and communication. All of these sectors reported the highest occupational accidents in the year 2016. In-depth studies on slip-and-fall may increase the awareness among workers with regards to these hazards at their workplace.